Y0-CJF-225

Date: 3/24/22

J 537.534 GAR
Garstecki, Julia,
Radio waves /

PALM BEACH COUNTY
LIBRARY SYSTEM
3650 SUMMIT BLVD.
WEST PALM BEACH, FL 33406

WAVES IN MOTION

RADIO WAVES

by Julia Garstecki

PEBBLE
a capstone imprint

Pebble Emerge is published by Pebble, an imprint of Capstone.
1710 Roe Crest Drive
North Mankato, Minnesota 56003
www.capstonepub.com

Copyright © 2021 by Capstone. All rights reserved. No part of this publication may be reproduced in whole or in part, or stored in a retrieval system, or transmitted in any form or by any means, electronic, mechanical, photocopying, recording, or otherwise, without written permission of the publisher.

Library of Congress Cataloging-in-Publication Data is available on the Library of Congress website.
ISBN: 978-1-9771-2273-5 (library binding)
ISBN: 978-1-9771-2618-4 (paperback)
ISBN: 978-1-9771-2300-8 (eBook PDF)

Summary: Radio waves aren't just for listening to your favorite radio station. Learn about radio waves and how we use them every day.

Image Credits
Capstone Studio: Karon Dubke, 17, 20; iStockphoto: peterhowell, 7; Shutterstock: 4 PM production, 9, Andrey_Popov, 11, Bors Rabtsevich, 16, DesignRage, 8, FTiare, 19, Graphic Compressor, 6, iDEAR Replay, 13, KENNY TONG, 15, Monkey Business Images, 5, V.Gordeev, Cover

Design Elements
Capstone; Shutterstock: Miloje, StockAppeal

Editorial Credits
Editor: Michelle Parkin; Designer: Ted Williams; Media Researcher: Jo Miller; Production Specialist: Laura Manthe

All internet sites appearing in back matter were available and accurate when this book was sent to press.

Printed and bound in China.
3322

TABLE OF CONTENTS

WHAT ARE RADIO WAVES? 4
WAVES AND WAVELENGTHS 6
HOW WAVES WORK 8
CARRYING VOICES 10
RADIO WAVES IN SPACE 14
STOPPING THE WAVES 18
BLOCKING RADIO WAVES 20
    GLOSSARY 22
    READ MORE 23
    INTERNET SITES 23
    INDEX 24

Words in **bold** are in the glossary.

WHAT ARE RADIO WAVES?

Sing along to the radio. Talk to a friend on the phone. Play a game on your tablet. Watch your favorite show on TV.

What do these devices have in common? They all use radio waves. Radio waves carry information from place to place.

5

WAVES AND WAVELENGTHS

We don't see radio waves. But they are all around us. Radio waves fill the air. They move like waves in water.

Radio waves can be short or long.
The wave's size is called the wavelength.
A radio wave's wavelength can be smaller than a pin. It can be longer than a city.

HOW WAVES WORK

Imagine playing a game of catch with your dad. He throws the ball. You catch it with a baseball glove.

Radio waves act like the ball. Instead of a glove, an **antenna** catches and throws the radio waves. It sends and catches information.

9

CARRYING VOICES

Watch someone speaking on your tablet. He or she talks into the tablet's **microphone**. It turns the speaker's voice into **energy**.

The energy moves to the tablet's antenna. The antenna turns the energy into radio waves. They travel through the air.

11

The radio waves carry the speaker's voice. The antenna in your tablet catches them. It turns the waves back into energy. The energy turns into words you can hear.

Wireless networks use radio waves too. **Wi-Fi** needs radio waves to work. Laptops, phones, and tablets have antennas. They catch the radio waves. Then they send the information to your screen.

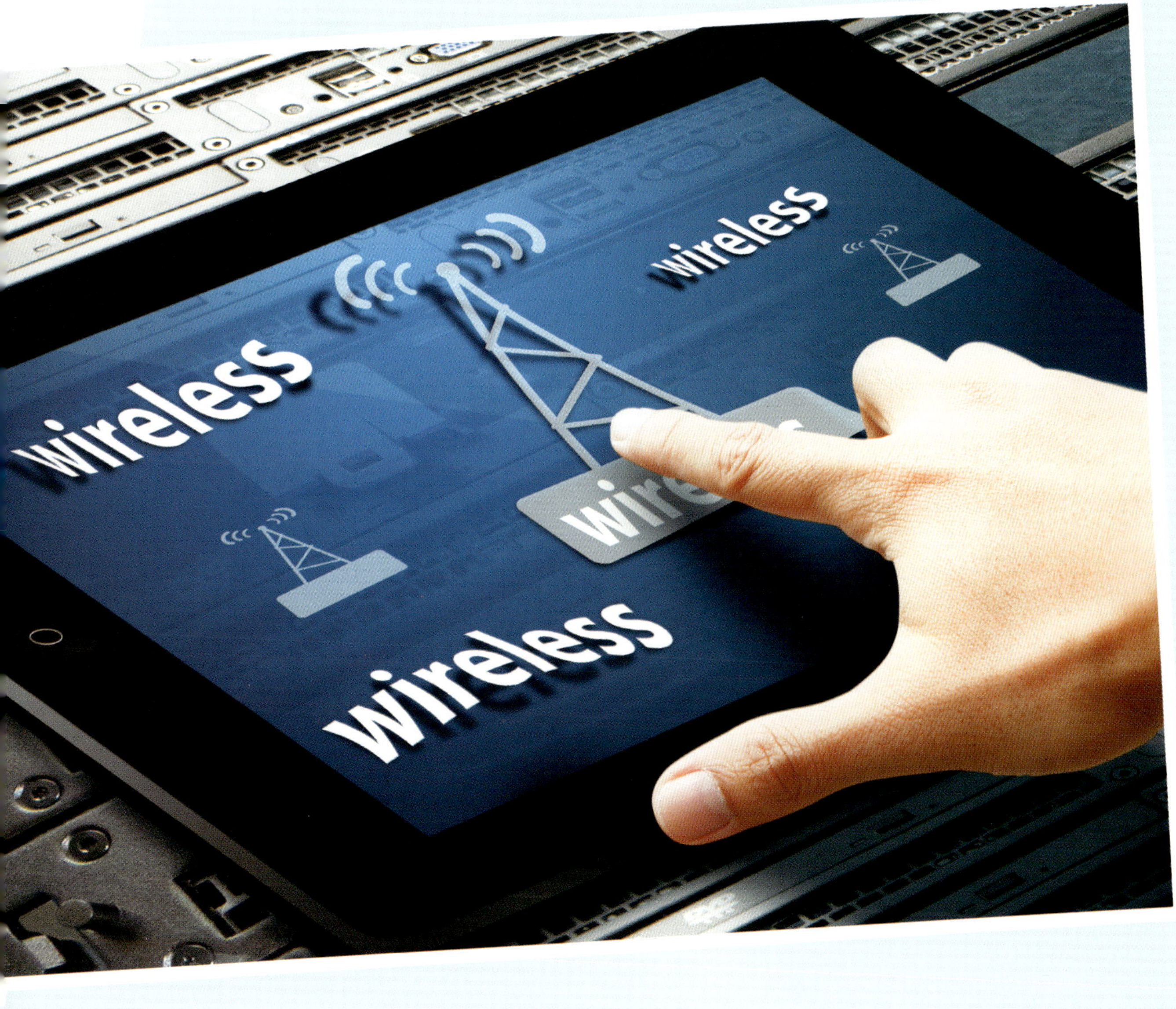
wireless
wireless
wireless

RADIO WAVES IN SPACE

Objects in space make radio waves. Scientists study these waves. Some spacecraft have special equipment. They record radio waves in space.

Scientists use **radio telescopes** to study waves in outer space. These big telescopes are on Earth. They tell scientists what these space objects are made of.

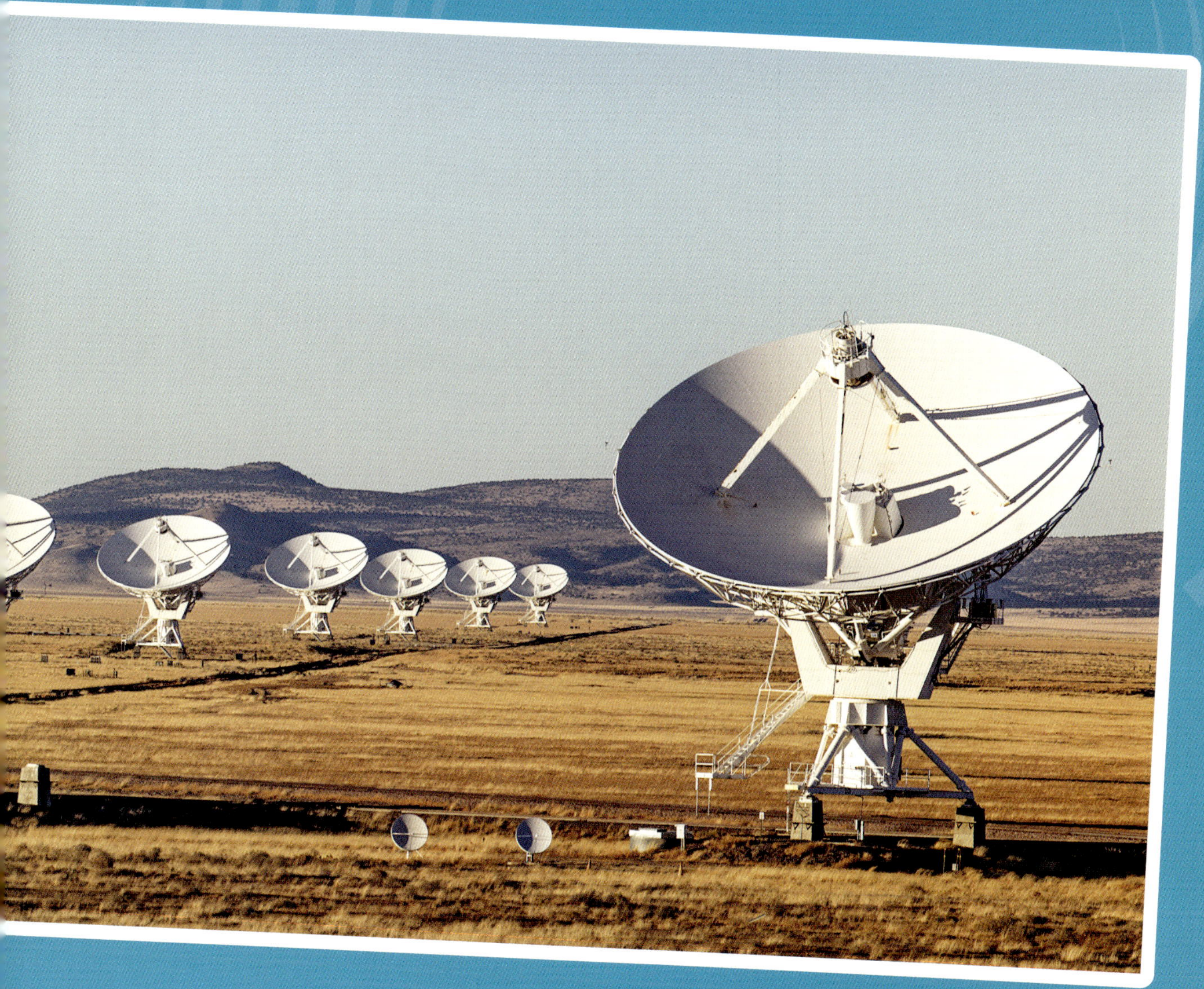

Many people use map apps on their phones to get to places. These apps use **GPS**.

GPS shows people where they are and where to go. It works with **satellites** in space. The satellites use radio waves to send information to GPS.

A satellite in space

17

STOPPING THE WAVES

You are listening to your favorite song on the radio. Your car goes through a tunnel. The radio gets quiet. What happened?

Radio waves can be blocked by objects, such as metal. Some places are harder for radio waves to reach. Our devices don't work in these areas.

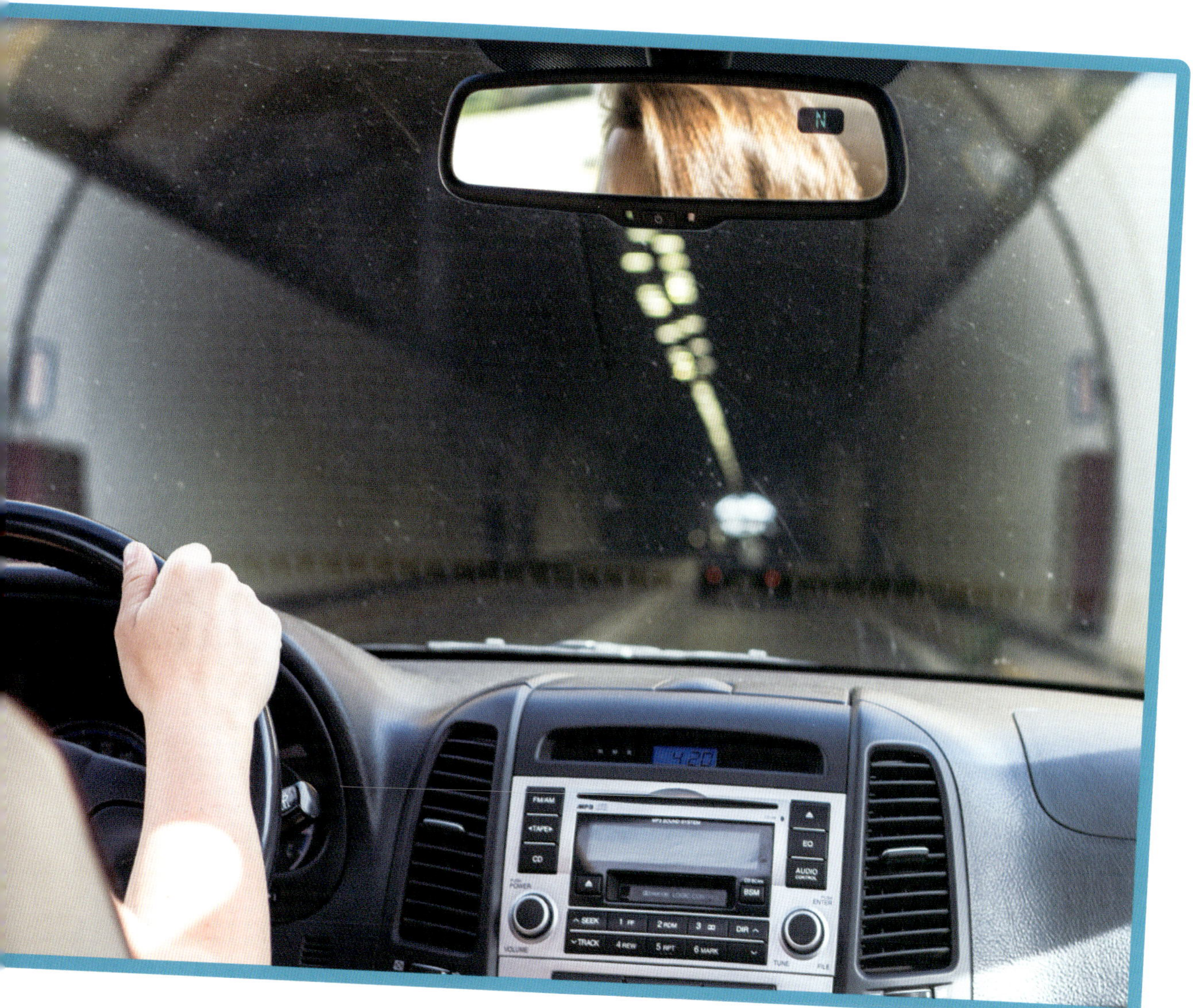
FM/AM
CD
EQ
AUDIO
BSM

BLOCKING RADIO WAVES

Radio waves tell a remote-control car to move. Find out what objects block radio waves.

What You Need:
- a toy car with a remote control
- a large piece of aluminum foil
- a large rubber glove
- a pillowcase
- plastic wrap

What You Do:
1. Drive your car around using the remote control.
2. Put the remote in the pillowcase. Make sure the remote is completely covered. Now try to drive the car. What happens?
3. Take the remote out of the pillowcase and wrap it in aluminum foil. Try to drive.
4. Try again using plastic wrap. Next, use the rubber glove.

   Can you still drive the car when the remote control is in the pillowcase? What about the other materials? Why or why not? Think of other materials to try.

GLOSSARY

antenna (an-TE-nuh)—a wire or dish that sends or receives radio waves

energy (E-nuhr-jee)—the ability to do work

GPS (GEE-PEE-ESS)—an electronic tool used to find the location of an object or place

microphone (MYE-kruh-fone)—a device used to make sounds louder, such as a person's voice

radio telescope (RAY-dee-oh TEL-uh-skope)—an instrument that collects radio waves put out by objects in space

satellite (SAT-uh-lite)—an object in space that circles Earth to gather and send information

Wi-Fi (WHY-FY)—a way to connect computers in the same area without using wires

READ MORE

Dahl, Michael. *Sound Waves*. North Mankato, MN: Capstone, 2021.

Diehn, Andi. *Waves: Physical Science for Kids*. White River Junction, VT: Nomad Press, 2018.

Dodd, Emily. *Energy*. NY: DK Publishing, 2018.

INTERNET SITES

Facts About Radio Waves
https://sciencewithkids.com/science-facts/facts-about-radio-waves.html

Physics for Kids: Types of Electromagnetic Waves
https://www.ducksters.com/science/physics/types_of_electromagnetic_waves.php

INDEX

antennas, 8, 10, 11, 12

apps, 16

blocking radio
    waves, 19

cell phones, 4, 12, 16

energy, 10, 11, 12

GPS, 16

information, 8, 12

laptops, 12

microphones, 10

radios, 4, 18

radio telescopes, 14

satellites, 16

space, 14, 16

spacecraft, 14

tablets, 4, 10, 12

TVs, 4

wavelengths, 7

Wi-Fi, 12